Double the Number
MW00592884

Dolphins have fins that
help them swim quickly.
They live in the ocean.

Can you count the fins the dolphin has **on each side**?

Spiders can be found in many places.
Some spiders make silk.
Most spin sticky webs to catch food.

Can you count the number of legs the spider has on one side?
How many legs does it have **in all**?

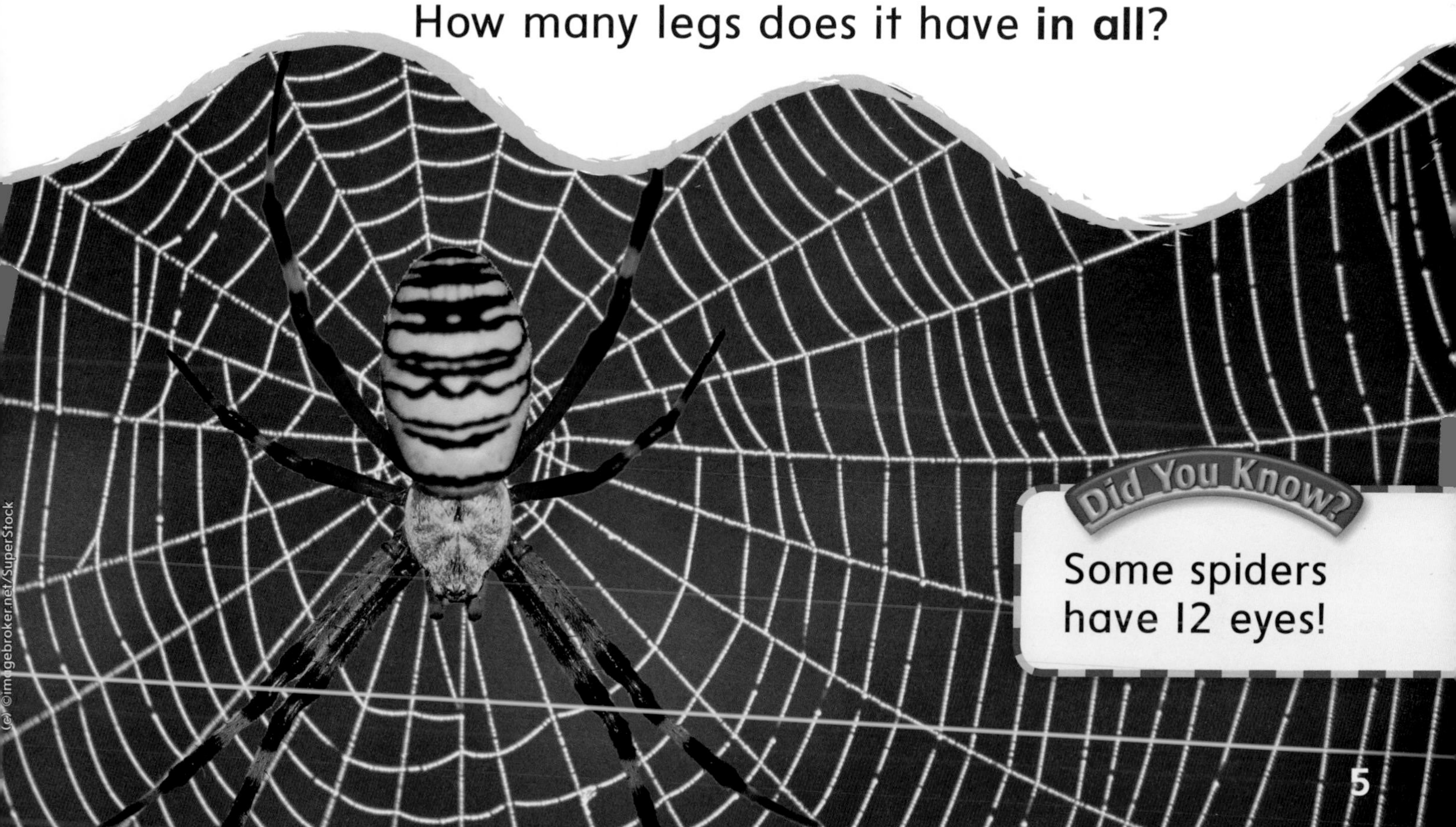

Did You Know?

Some spiders have 12 eyes!

It is very cold where polar bears live.
They have thick fur and padding that keeps them warm.
They have webbed paws for swimming and sharp claws for catching fish.

Can you count the number of pads the polar bear has on one paw?
How many pads does it have on all its paws?

Geckos are reptiles that are good diggers.
Can you count how many toes the gecko has on each foot?
How many toes does a gecko have **in all**?